U0178731

粤雅小丛书

ElegantGuangdong Series

Gambiered Canton Gauze

香云纱

云纱霓裳

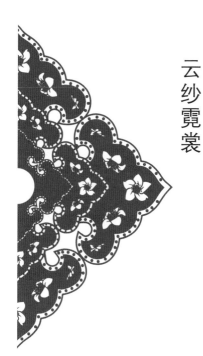

南方日报
出版社
NANFANG DAILY PRESS

· 广 州 ·

粤雅小丛书编委会

编

图书在版编目（CIP）数据

香云纱：云纱霓裳 / 粤雅小丛书编委会编. — 广州：南方日报出版社，2020.12
（粤雅小丛书）
ISBN 978-7-5491-2263-9

Ⅰ. ①香… Ⅱ. ①粤… Ⅲ. ①绞纱－染整－介绍－广东 Ⅳ. ①TS190.63

中国版本图书馆 CIP 数据核字(2020)第 221594 号

XIANGYUNSHA：YUNSHA-NISHANG
香云纱：云纱霓裳

编　　　者：	粤雅小丛书编委会
出版发行：	南方日报出版社
地　　　址：	广州市广州大道中 289 号
出 版 人：	周山丹
责任编辑：	阮清钰　陈　静　黄敏虹
翻　　　译：	吴忠岫
责任技编：	王　兰
责任校对：	魏智宏
内文设计：	邓晓童
封面设计：	书窗设计
经　　　销：	全国新华书店
印　　　刷：	广州市岭美文化科技有限公司
开　　　本：	787mm×1092mm　1/32
印　　　张：	2.125
字　　　数：	30 千字
版　　　次：	2020 年 12 月第 1 版
印　　　次：	2020 年 12 月第 1 次印刷
定　　　价：	25.00 元

投稿热线：(020) 87360640　　读者热线：(020) 87363865
发现印装质量问题，影响阅读，请与承印厂联系调换。

【目录】

【引言】

    "荔熟蝉鸣云纱响，蔗浪蕉风莨绸爽。"莨绸就是广东珠三角地区生产的一种古老的染色面料香云纱。香云纱透气清爽，轻体透凉，穿着它走路会发出"沙沙"的声音，所以最初被称作"响云纱"。

    繁复的制作工艺以及特殊的丝绸质感，使香云纱享有"软黄金"的美誉，历史上曾卖出过十二两白银一匹的天价，是中国丝绸界不折不扣的"黑富美"。香云纱区别于一般丝绸布料的地方，是它如美酒醇酿，穿得越久越有魅力。香云纱布料与身体发生化学反应，会慢慢现出褐黄色的底色，布料表面的纹理会越来越细，越穿色泽越油润乌亮，仿佛有生命一般。

香云纱的"铁粉"很多，最出名的是"宋氏三姐妹"。上海宋庆龄纪念馆里收藏了宋庆龄最爱的黑色香云纱旗袍。这件旗袍陪她出席过许多公开活动，她在老年身体发福后，还将香云纱旗袍加宽了继续穿，喜爱之情可见一斑。20世纪20年代上海名媛，如张爱玲、陆小曼、王映霞等，也都是香云纱的追捧者。张爱玲不仅让小说里的人物穿上香云纱，在为自己的英文书《中国人的生活和时装》配的时装图里，她也选了香云纱衣服的图片。由于香云纱布料的珍贵稀有，据说以前广东女子出嫁时，母亲都要亲手缝制一套香云纱衣服送给女儿，等女儿老了，又作为传家宝送给下一代。

香云纱的出现，是广东两千多年丝织史上一块重要的里程碑，创造了广东丝织工业史和丝绸贸易史上的辉煌。2008年香云纱因独特的染整技艺被列入国家级非物质文化遗产。从原料到制作都是纯天然的特性，在越来越注重环保理念和传统技艺传承的今天，香云纱重新得到关注和青睐。

第一章

催生演化

——香云纱的起源

源头：桑基鱼塘　奠定基础

《中国大百科全书·纺织卷》里对香云纱的解释是：表面乌黑光滑有透孔小花的丝造物（莨纱），"经煮练、脱胶、上莨、过乌、水洗等加工而成"。2008年《香云纱染整技艺申报国家第二批非物质文化遗产名录材料》里则认为，香云纱是蚕丝平纹织物（绸）和蚕丝纱罗组织织物（纱）经过染整后的产品。虽然概念的范围有大小，但是有两点一致：第一，香云纱的原料是丝绸；第二，它是经过染整技艺加工的制品。

种桑、养蚕、缫丝和织绸在珠江三角洲是一项传统而古老的产业，有2000多年的历史。

香云纱原料：丝绸

珠三角地区河涌交织的特殊环境，产生了一种独特的农业生态链"桑基鱼塘"。这是当地人的智慧创造，在池塘埂上种植桑树养蚕，蚕粪和蚕蛹可以作饵料喂鱼，塘泥又成为桑树的肥料，循环往复。由于桑蚕丝的收益巨大，历史上从明朝到清朝，珠江三角洲发生过三次弃田筑塘、废稻树桑的热潮，造就发达的桑蚕生产基地，由此带动了当地缫丝和织绸等加工产业的进步。

在太阳底下整齐晾晒的香云纱

在明朝，广东生产的丝绸产品已经全国闻名。乾隆《广州府志》记载："粤纱，金陵、苏、杭皆不及。"但那时候纺织原料都是采用苏杭一带的蚕丝，因为吴丝质量高，织出的丝绸较之本地丝色彩好、光泽度高。自1840年鸦片战争后，清政府新开放上海等五处通商口岸，打破广州"一口通商"的地位。苏杭蚕丝不再通过广州而选择邻近的上海出口。尤其是太平天国运动爆发后，各地物流交通被切断，然而国际蚕丝需求依然很大，广东的厂家只能依靠本地土丝生产丝绸产品，也因此促进珠三角桑蚕业快速发展。

初遇：薯莨纱绸　见泥则黑

岭南山里寻常见的薯莨作为优良的红褐色植物染料，很早就被岭南人们使用。北宋沈括的《梦溪笔谈》记载岭南人用薯莨来染皮制靴子。之后的文献也能查到薯莨被当做染鱼网、绳索、布料的染料用。但薯莨应用到丝绸面料的染色上的明确记载，最早要到清宣统三年（1911年）的《番禺县续志》："市桥人多种薯莨为业，用以染绸曰薯莨绸""所出薯莨纱绸名驰远近"。薯莨纱绸，就是经过晒莨后的丝绸产品，这说明在清代已经有广义上的香云纱出现。

香云纱经薯莨浸染后晾晒

薯莨染过的丝绸面料是红褐色的，只有在涂抹过河泥，河泥中的铁元素与薯莨汁发生微妙反应后，丝绸涂过河泥的那一面才会形成类似树脂的黑色纯天然薄膜，这是染整技艺中的关键步骤——过乌。有研究者认为，过乌源于渔民晾晒渔网时发现渔网变黑而得到的灵感。

珠三角一带的渔民爱用薯莨染渔网，使渔网有韧性，耐潮不易腐蚀。屈大均所作《广东新语》记载："薯莨胶液本红，见水则黑。诸鱼属火而喜水，水之色黑，故与鱼性相得。"根据这个记载，研究者分析，见水则黑，是因为渔网触碰到河底淤泥，渔网出水的时候，淤泥又被洗掉，这个过程犹如过乌及洗泥。捕完鱼，渔网在阳光下晾晒，泥中的铁与渔网中的薯莨汁发生反应，使渔网最终变成黑色。通过对渔网颜色变化的观察研究，人们把过乌这个染整技艺被运用到了香云纱上，使其正面呈现为富有亮度与光泽的黑色。

第二章

天人合一
——香云纱的染整过程

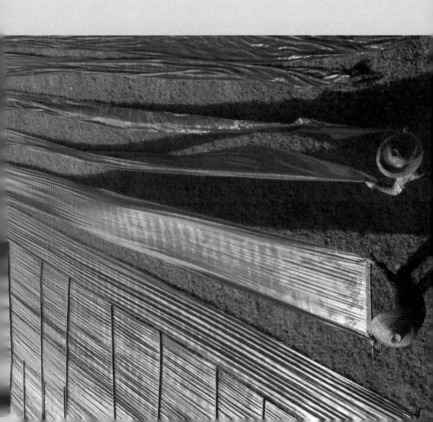

取材：本自天成　因地取材

　　香云纱的染整过程，沿袭了几百年以前的古老工艺要求。三个关键要素——薯莨汁、河泥和阳光，无一不是取之自然；作为坯底的丝绸，也是选了蚕丝纺织而成。可以说，香云纱是大自然恩赐给岭南的一件艺术珍品。

　　坯底丝绸根据组织和花纹的不同，分为"白坯绸"和"白坯纱"。"白胚绸"是平纹组织丝织物，又叫"生绸"，组织细密平整；"纱"是民国初年南海工匠改良提花机织，生产的具有扭眼通花图案的绞纱丝绸织物，俗称"白坯纱"。这种白坯纱面料上有许多透气孔，更加适合潮湿闷热的广东。

制香云纱的关键原料——薯莨

　　薯莨是生长在亚热带地区深山中的藤本植物的块茎，在两广地区最为常见。它的块茎表皮呈黑褐色，坑洼不平。如将其切开，会发现里面是暗红色或红黄色。薯莨在中药里应用广泛，可以用来止血、止痛和抗菌。它的块茎汁液富含单宁，是香云纱的基础染色材料。

　　长期以来，香云纱只在珠江三角洲的顺德、南海等地生产，就是因为制作香云纱另一样必不可缺的材料河泥并非一般的河泥，而必须是没有被污染过的含铁河泥。这样的河泥，只有珠三角地区的河流才有。这些地区的河泥没有被污染，十分润滑，所以日本等国都有尝试要在本国进行香云纱生产，均因为香云纱布料形成乌黑亮泽涂层的关键材料河泥不可得，而纷纷以失败告终。

制作香云纱是个靠天吃饭的活。浸染薯莨汁的丝绸必须通过太阳反复曝晒，来使薯莨汁中的单宁充分渗透到白纱绸胚里面，在布料表面沉积形成涂层。如果晒莨时遇到下雨，雨水会使一匹布的上色不均匀。广东的天气，只有4月至10月有充足持久的阳光，适合晒莨。7月和8月的日照太强，气温过高，这样的天气晒出来的纱绸容易变硬发脆，不适合开工。到了11月以后，北方干燥的季候风南下，同样不适合晒莨。所以一年内真正可以晒莨的日子，不过4个月。

步骤：匠人匠心　因材施艺

香云纱的染整技艺又称"晒莨"，有14道主要工序，包括坯绸准备、浸薯莨水、晒莨、封莨水、煮炼、过河泥、水洗、拉布、摊雾、卷绸、整装入库等，整个流程十分复杂烦琐，完全依靠手工，一般需要15天左右才能完成全过程。无论是薯莨水的浓度，还是晒莨、封莨的次数，每一道工序都只能靠染整师傅凭借经验，根据具体情况及时灵活地进行调整，"因天而异、因地而异、因场而异、因人而异"，制作了每一匹独一无二的香云纱。

香云纱的染整有以下关键几步：

一是调制薯莨水。薯莨磨碎，浸在槽里产生薯莨汁，过滤掉薯莨渣，把薯莨汁放进大木桶待用。不同品质、不同织法的丝绸对薯莨汁的浓度要求不一样。薯莨汁调制的好坏，是决定香云纱品质的关键。薯莨汁的浓度没有标准，负责调制的老师傅们，按照每天要染整的

浸染薯莨汁

布料数量，以及染整的速度随时调节薯莨汁浓度。其间的比例配置，师傅们只能依靠自己长年积累的经验，没有书本可以参考。

二是浸泡晒制。绸匹裁成20米一段（方便晒莨工人一人操作），放入装了薯莨汁的水槽中。老师傅开始不断地用手翻动绸匹，目的是为了使绸匹完全浸透吸匀薯莨水。随后，从浸槽取出来的绸匹经过自然脱水后，再由工人沿草地平整拉直，曝晒至干。为了使染色更加均匀，绸匹晒干后，工人会用桶盛薯莨水喷洒到绸匹上，之后再用蒲叶帚扫匀。曝晒至干、浸染、再曝晒，这个过程需要反复多次。

在草地上晾晒的作用，是高温曝晒下绸匹可从草地中吸取一定量的水分，同时保持绸匹清洁，不会让丝绸与泥地细沙直接接触。而且绸匹阴阳面温差较大，可增加香云纱"阴阳色"效果。因此，草地既要柔软，不会划伤丝绸布料，又要能承托一定的重量，所以为晒香云纱培育的草地一般选择1—2厘米厚，俗称

"爬地老鼠"的本地草种为宜。

三是过乌，即涂河泥。这道工序是晒莨的关键，取珠三角河涌里特有的河泥，均匀搅拌成糊状，用特制的拖把均匀涂在绸匹的正面。

工人抱着一大叠香云纱去阳光下的草地上晾晒

　　涂好河泥的绸再由操作人员两人一组，平稳拉直抬到草地上，静候半小时。这个过程可以保证河泥中的高价铁离子与薯莨中的单宁产生充分的化学反应，生成黑色沉淀物凝结于绸缎表

过乌，是"软黄金"香云纱最特别的工艺

面。为了使河泥与绸匹有充分的时间进行化学
反应，在没有大棚遮挡之前，过乌只能在日出
之前完成。

绸匹过乌后，需要立刻运到河涌里，洗掉
布料上剩余的河泥，接着绸匹会重新平摊到草
地上，此时太阳刚刚升起，清晨的阳光把绸匹
晒干。至此，涂抹河泥的绸匹正面已呈现出乌

黑油润的外观，增添了如陶瓷一般的光泽，而反面由于没有接触河泥，依旧为红棕色。

四是摊雾。傍晚太阳下山以后，草地沾满露珠，此时，过泥曝晒后的硬硬的莨绸被重新铺在草地上，吸收草地上的水分，绸匹渐渐软

过乌后，还需清洗绸坯上的河泥

化。这个过程叫做"摊雾"。

　　摊雾是晒莨的最后一道工序，但是此时的
香云纱并非最终成品，经过卷绸、码尺、整装

通常，过乌都在夜晚进行

入库，在仓库存放3—6个月后，还要进行下一步操作。

第三章

破茧成蝶
——香云纱的发展困境和展望

低谷：传承凋零　后继乏力

结合天时、地利、人和的纯手工制作工艺造就了香云纱的不可复制性，也成为香云纱的传承瓶颈。传承人梁珠老先生介绍，香云纱制作技艺对环境的依赖需要传承人非常细心，并且有极高的关注度，出一点差错便前功尽弃。

香云纱的制作要经受长期日晒，劳动强度大。以使香云纱阴阳两面获得最理想效果的最原始、最古老环节"晒莨"为例，每年的4月初到10月底，是岭南最热的时节，却也是晒莨的最好时间。每天从凌晨3时到下午4时，工人需要忍受疲惫、暴晒，不断重复晾晒的过程，

香云纱传承人梁珠老先生

铺展、拉直、喷水、收敛，为了使绸匹的染色均匀，控制阳光照射时长，他们中间几乎没有停歇，工作强度非常大。因此，当代真正懂得制作香云纱绝技的人很少，且年龄偏大，年轻人都不愿意学，后人能否延续香云纱的神话，是目前面临的一个非常严峻的问题。

另一个困境是市场需求问题。20世纪30年代，是香云纱生产的鼎盛时期，生产规模很

大，工人人数上万。但是，随着市场替代品，
如化纤和织品等的兴起，香云纱产量逐年下
降。在 20 世纪 80 年代，不少香云纱工厂倒闭
转产。目前，全国专业生产香云纱的厂家仅有
3 家，且主要面向国外市场。

香云纱·云纱霓裳

突破：工序改良　时尚助力

　　年逾八十的梁珠老先生从十四岁在香云纱厂当学徒，经历了香云纱发展的起起落落。20世纪90年代，为了不让香云纱工艺失传，他买下倒闭的香云纱工厂，凭借经验积累，在传统工艺的基础上，改进晒莨流程，形成一套"三洗九煮十八晒"的工序，逐步赢得市场认可，获得许多海外订单。

　　经过顺德伦教政府部门的大力扶持，2007年"香云纱晒莨工艺"即"香云纱染整技艺"成功申报为省级非物质文化遗产，2008年晋级为国家级第二批非物质文化遗产。更有年

精美的香云纱纹饰

轻人拜梁珠先生为师，愿意学习香云纱染整技术，传承衣钵。

中国十佳服装设计师、岭南服饰博物馆馆长及广绣传承人、顺德籍设计师屈汀南从小在晒莨厂旁边长大，对香云纱有深厚的感情，一直致力于发展与传承香云纱工艺。他以莨纱为布料设计作品参加联合国组织的服装设计大赛，捧回"世界民俗文化奖"，得到国际专家的肯定，也让世界重新认识了香云纱。

工艺改良后，质感与美感、时尚感兼具的香云纱

　　不仅如此，屈汀南还请专家研制色彩多样的莨纱，并成功研制出红色的熟莨。2018中国国际时装周，屈汀南在"香祥响向"发布会上，用色彩缤纷的香云纱面料制作高级时装，让中国非遗文化艺术香云纱再次隆重登上世界舞台。

　　香云纱，这个由几代人甚至十几代人的传承与发展而形成的优美织物，作为一张历久弥新的对外交流名片，在重视传统、回归自然、

香云纱T台秀

2016年时尚管理EMBA学术交流会现场展出的香云纱作品

崇尚环保的今天，以其自然天成的特性，走
出了一条文化的本土性和时尚的世界性完美
结合的设计之路，越来越受全国乃至世界人
们的喜爱。

Gambiered Canton Gauze:
Flowery Silk on the Wing

CONTENTS

Foreword

"When litchi ripens, cicada chirps, the gambiered Canton gauze rustles; sugarcanes roll, plantain leaves ruffle, in gambiered gauze I luxuriates."

Gambiered Canton gauze is an ancient species of dyed gauze produced in the Pearl River Delta area. The gauze is of permeable, refreshing and light fabric, which brings about cooling effect to the wearer. As the gauze gives off the "rustling" sound in walking, it was originally named "rustling gauze".

Complicated production process and special silk texture brings about the repute of "soft gold" to gambiered Canton gauze. History has seen the exorbitant price of 12 taels of silver trading for a bolt of gauze. It is the worthy peak of "perfection" in the realm of silk. The gauze differs from other silks in that it mellows like fine wine over time, the more it wears the more charming it turns. Contact with skin will induce a chemical reaction in the fabric of the gauze; it will slowly develop isabellinus grounding. The longer it wears the finer the grain on the surface will develop, the more oily and black it will show, as though it has come alive.

The gauze has drawn many die-hard fans, the Soong Sisters were among the best known. The Soong Ching Ling Memorial Hall in Shanghai houses Soong Ching Ling's favorite black cheongsam made of gambiered Canton gauze. This cheongsam had been with her on many public occasions. When she gained weight as she aged, Soong Ching Ling had the cheongsam let out for continued wear. Her deep love for the cloth could be seen therefrom. In the 1920s, many socialites including Eileen Chang, Lu Xiaoman, Wang Yingxia were all fans of the gauze. Eileen Chang not only put the gauze on the her novel characters, but also chose pictures of gambiered Canton gauze as accompanying illustrations of her English book *Chinese Life and Fashions*. Due to rarity of the fabric, it is said that a mother would hand sew a set of gauze clothes as dowry for her daughter at her marriage. When the daughter gets old, she would then bequeath the gauze as a family heirloom to her offspring.

The appearance of gambiered Canton gauze marked an important milestone in the history of China's two-millennia long silk product development, which brought about the glory attained in the histories of silk industry and silk trade of Guangdong province. In 2008, the gambiered Canton gauze was placed into the catalog of national intangible cultural heritage for its unique dyeing and finishing technology. At a time where people are paying growing attention to ideas of environment protection and preservation of traditional technology, the characteristics of naturalness from the source of materials to production has won back customers' attention and favor.

Chapter I

Birth and Evolution—Origin of Gambiered Canton Gauze

Source: Mulberry-Fish Ecosystem Laying Foundation for the Birth of the Gauze

The Encyclopedia of China—Volume of Textile defines the gambiered Canton gauze as glossy and sleek jet-black silk fabric (dye yam gauze) with open-work florets, "the fabric is made with procedures including scouring, degumming, applying extracted dye yam solution, applying dark river mud, rinsing". In 2008, *Application Materials of Gambiered Canton Gauze for Listing in the 2nd Batch of National Intangible Cultural Heritage Catalog* holds that gambiered Canton gauze is a plain fabric silk product (silk) and a silk gauze fabric (gauze) after dyeing and finishing processes. There are differences in the scope of the two concepts; however , the two concepts have two points in agreement: first, it is made of silk; second, it is a product processed with dyeing and finishing techniques.

Growing mulberry, sericulture, silk reeling and silk weaving are traditional industries with a long history of over two thousand years in the Pearl River Delta. The special environment of interlacing rivers in the Pearl Delta River area gives rise to a unique agricultural ecological chain, the "mulberry-fish ecosystem". It is a work of wisdom by the local people. They grow mulberry and raise silkworm on fish pond banks, the silkworm feces and silkworm can feed fish, while pond mud turns further into fertilizer to fertilize the mulberry, hence a repeatable cycle is formed. Due to the tremendous profit obtained from raising silkworm and weaving silk, during the period from the Ming Dynasty to the Qing Dynasty, there occurred three rushes of building fish ponds by destroying crop land, planting mulberry by scraping rice paddies in Pearl River Delta. These rushes brought about well-developed mulberry and silkworm production bases, which in turn ushered in the progress of industries such as silk reeling and silk weaving.

By the Ming Dynasty, silk products from Guangdong had already made their name across the nation. According to *Annals of Guangzhou Prefecture* in the reign of emperor Qianlong, "Canton silk is unrivaled by that of Nanjing, Suzhou and Hangzhou." However, raw silk for the weaving industry were mainly shipped in from the Suzhou-Hangzhou area, as silk from the place was of better quality, and silk fabric weaved with such silk superior in color and luster to that weaved with locally purchased silk.

Since the Opium War of 1840, the Qing government opened five additional trading ports including Shanghai, which practically shattered Guangzhou's status as the "sole trading port". Export of silk produced in the Suzhou-Hangzhou area was no longer conducted via Guangzhou, instead Shanghai in close vicinity was chosen as the export port for such silk. After the breakout of the Taiping Heavenly Movement, trade routes between the various places were cut off, however, international demand for silk remained strong. Guangdong merchants had to count on local silk producers for the production of silk fabric, which facilitated the rapid growth of mulberry and silk industries.

First Encounter: Yam Juice Dyed Gauze Blackening When Exposed to Mud

As a superior plant dyeing material for bronzing, the commonly available dye yam in the mountainous areas south of the Five Ridges had been put to use by the local people. In his works *Dream Creek Essay*, Shen Kuo of the North Song Dynasty recorded the local practice of people living south of the Five Ridges dyeing leather for shoes with dye yam. Practices of dyeing fishnet, rope, cloth with dye yam can be found in literature thereafter. However, the earliest definite record of employing dye yam as a silk fabric dye material appeared in *The County Annals of Panyu* in the third year of the Xuantong period (1911), "the people of Shiqiao town mostly grow dye yam as a means of livelihood, which is used as a dye material in the making of dye yam gauze", "the dye yam gauze produced therefrom made its name known far and wide". The so-called dye yam gauze is the end silk product resulting from applying extracted dye yam solution. The fact indicates that there had already been gambiered Canton gauze in the broad sense in the Qing Dynasty.

The silk fabric appears bronzing after dyeing process with dye yam. Apply river mud and let the iron element undergo a delicate reaction with the dye yam solution, only after this process can a resin-like purely natural black membrane be formed on the side whereon river mud is applied. This is the critical step in the entire dyeing and finishing process—mud application. Some researchers hold that, the mud application process is inspired by the fact that fishermen discovered that fishnet turns black when it is taken from water and aired.

007

Fishermen in the Pearl river delta are fond of dyeing fishnet with dye yam to make it more resilient, moisture-resistant and corrosion-resistant. Qu Dajun recorded in *New Remarks on Guangdong* the following, "the dye yam solution is red, it turns black when it comes in contact with water. The fish pertains to the virtue of fire, hence it favors water; the water is of black color, hence it accords with the nature of the fish." Researchers analyzed the phenomenon according to this entry and proposed that, the fact fishnet turns black when coming into contact with water is that fishnet comes into contact with mud on riverbed and mud is rinsed off when the fishnet is pulled out of water. The process is like applying and rinsing mud. Having finished a fishing trip, the fishermen air dried their fishnets in the sun, iron in the mud reacted with the dye yam solution in the fishnet, the fishnet finally turned black. Through observation and analysis of the color change of the fishnet in the process, people invented the mud application technique and applied the dyeing and finishing technique to production of gambiered Canton gauze, which turns the front side of the fabric into glossy and lustrous black.

$\mathcal{C}hapter\ \mathcal{II}$

Union of Heaven and Man—the Dyeing and Finishing Process

Material: Naturally Endowed and Locally Sourced

The dyeing and finishing process of gambiered Canton gauze follows ancient technological requirements handed down for hundreds of years. The three key factors include dye yam solution, river mud and sunlight, none of which is not naturally obtainable. The silk fabric ground on which gauze is made is woven with silk produced by silkworm. It is justifiable to say the gambiered Canton gauze is an artistic treasure endowed to the people living south of the Five Ridges by nature.

The silk greige could be classified into two categories: "the white silk greige" and "the white gauze greige" according to the texture and grain of silk fabric ground. "The white silk greige" refers to plain weave texture silk products, or raw silk, it is of fine and even texture; "guaze", folk name for which is "the white gauze greige", refers to skein silk products woven with jacquard which had been innovated by craftsmen from Nanhai County in the early years of the Republic of China, featuring fretwork of jacquard weave. There are myriad pores on the fabric, which makes it all the more fit for wear in the humid and sultry weather of Guangdong area.

Dye yam is the tuber of a vine plant that grows in the remote mountains of subtropical regions, and is most commonly seen in Guangdong and Guangxi. Its tuber skin is of dark brown color and bumpy appearance. If you cut it open, you will find that the inside is dark red or reddish yellow. The yam is widely used in traditional Chinese medicine. It can be used to stop bleeding, relieve pain and sterilize bacterial. Its tuber juice is rich in tannins, which is the basic dyeing material for gambiered Canton gauze.

For a long time, the gambiered Canton gauze could only be produced in Shunde, Nanhai and other places in the Pearl River Delta area, because there is another indispensable material for making the gambiered Canton gauze which is not the ordinary river mud, but unpolluted river mud with iron element presence. This kind of mud is only available in rivers of the Pearl River Delta area. The river mud in these areas has not been polluted and is very fine and smooth. Therefore, although Japan and other countries have tried to produce gambiered Canton gauze in their own countries, they have all ended up in failure due to the fact that the river mud, the key material for forming the black and shiny coating on the gauze, cannot be replicated.

Making gambiered Canton gauze is a craft that depends on the elements. The silk impregnated with dye yam solution must be repeatedly exposed to the sun to allow the tannins in the yam juice to fully penetrate the white gauze greige and deposit a coating on the surface of the cloth. If it rains while drying, the rain will produce a piece of cloth unevenly colored. Only April to October are the time when the weather of Guangdong is fit for drying the gauze with long and adequate sunlight. In July and August, the sunshine is too strong and the temperature too high; in such weather, the gauze dried tends to be rigid and brittle, which is not the time ideal to commence production. While in November, the dry northern monsoon wind blows southward, it is also not suitable for drying the gauze. So the time fit for drying the gauze is no more than four months in a year.

Procedures: Craftsmanship and Ingenuity

The dyeing and finishing technique of the gambiered Canton gauze is also called "gauze drying". It consists of 14 main processes, including preparation of silk fabric ground, soaking with dye yam solution, drying, sealing solution, boiling and extracting, applying river mud, rinsing, stretching, spreading, rolling, packing and warehousing. The entire process is very complicated and cumbersome, relying entirely on manual work. it generally takes about 15 days to complete the whole process. Regardless it be the concentration of dye yam solution, or the number of times of drying and sealing, each process can only be adjusted by the skilled craftsmen with reference to experience. The craftsman makes timely adjustments according to the specific situation, "varying adjustments on different weathers, different places, different workshops, different handling personnel", which makes every bolt of gambiered Canton gauze unique in its own way.

There are key steps in the dyeing and finishing process of gambiered Canton gauze as follows:

The first step is preparing dye yam solution. Grind the yam, immerse it in a trough to extract yam juice, filter out the yam dregs, and pour the solution in a large wooden barrel for later use. Different qualities and different weaving techniques have different requirements for the concentration of yam solution. The quality of prepared yam solution is the key to determining the quality of gambiered Canton gauze. There is no established standard for the concentration of the dye yam solution. The craftsmen in charge of preparation adjust the concentration according to the quantity of fabric to be dyed and finished and the speed of dyeing and finishing every day as they see fit. The concentration for a specific use relies exclusively on the experience of the craftsmen accumulated over the years, and there is no guidebook to refer to.

The second step is soaking and drying. A bolt of silk is cut into 20-meter pieces (fit for handling by one person), and placed in a sink filled with yam solution. The craftsmen begins to stir continuously the silk with his hands to make the silk fully soaked and absorb the yam solution evenly. The silk is then taken out of the tank and aired naturally; then the silk is unfolded and straightened out on the lawn, exposed to the sun until dry.

013

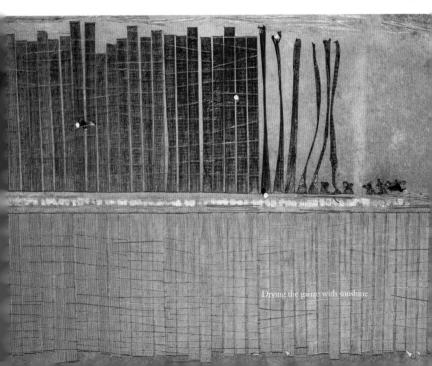

Drying the gauze with sunshine

After the silk is dried, in order to make the dye more evenly, the workers will spray yam solution on the silk from a bucket. The sprayed yam solution will then be swept with a cattail broom to spread it evenly on the silk fabric. The process of drying, soaking, and redrying will be repeated many times.

The point of drying silk on the grass is that the silk can absorb a certain amount of moisture from the grass when exposed to high temperature, while the silk can be kept clean when it is kept away from soil and sand by the grass beneath. While the substantial temperature difference between the front side and the back side of silk can enhance the "sharp-dull" contrastive effect of gambiered Canton gauze. Therefore, grass for such purpose must be soft, which would not scratch the silk fabric, while it can support a certain weight. Therefore, the grass cultivated for drying gambiered Canton gauze is generally trimmed to 1—2 centimeters high. The local grass species with the folk name of "ground-crawling mouse" is fit for this purpose.

The third step is applying river mud. This process is the key to silk drying process. Extract river mud unique to the rivers of the Pearl river delta area, stir it into a evenly mixed paste, apply the paste evenly on the front side of the silk with a specially made mop. The silk covered with river mud is then lifted in tightly stretched state to the grass by two workers. The silk then sits still for half an hour. This process can ensure that the high-valent

A worker is spreading the gauze in the sun

iron ions in the river mud and the tannins in the yam solution induce sufficient chemical reaction to produce black precipitates that condense on the front side of the fabric. In eras prior to the creation of greenhouses which can provide shade and shelter, in order to allow sufficient time for the chemical reaction to take place between river mud and the silk, the "mud application" process had to be completed before sunrise.

After the mud application process, the silk must be moved to the river immediately to wash off the remaining river mud on the fabric, and then it will be spread flat on the grass again. At this time, the sun has just risen and the early morning sun can dry the silk. By this time, the front side of the silk covered with river mud displays a black and oily appearance, adding a pottery-like luster, while the back side still remains reddish brown because it has not been exposed to the river mud.

The fourth step is to spread out the silk to absorb fog moisture. After the sun goes down in the evening, dew condenses and forms on the grass. At this time, the rigid silk fabric after mud application and drying is spread again on the grass to absorb moisture from the grass. The silk gradually softens. This process is called "spreading out for fog moisture absorption".

Fog moisture absorption is the last process; however, the gambiered Canton gauze being processed thus far is still not yet the final product. After steps of rolling, measuring, packing and committing to warehouse, it will then sit in the warehouse for 3— 6 months before being moved for the next step process.

Chapter *III*

A Reborn Artifact—Development Dilemma and Prospect

Slump: Dwindling Enthusiasm in the Transmission of Tradition

The entirely hand-made craftsmanship combined with time, location, and people have made gambiered Canton gauze unreplicable, and which has also made it the bottleneck in gambiered Canton gauze inheritance. Mr. Liang Zhu, the inheritor of the craft, explained that the reliance of skills in making gambiered Canton gauze on the elements requires the inheritor to be very careful and demands a high degree of attention, a little misstep would jeopardize the entire process.

Long time exposure to the sun in the production of gambiered Canton gauze entails intensive hard labor. Take the most primitive and most ancient practice of "drying gauze" for instance, in order to effect the optimal sharp-dull contrast effect between the front side and the back side, workers had to work from 3 a.m.

to 4 p.m. during the period from the beginning of April to the end of October each year—the hottest season in places south of the Five Ridges and yet the best time for drying gauze. Workers had to endure fatigue, exhaustion, scorching weather, and had to repeat the airing and drying process. In order to obtain even dyeing effect of the silk, workers had to work with almost no break throughout the day, carrying out spreading, straightening, spraying water, and contracting processes. Therefore, there are very few people who have really mastered the know-how in making gambiered Canton gauze in modern times, and those who have mastered tend to be old; while young people are unwilling

A fashion show of gambiered Canton gauze

to learn the trade. Whether future generations can inherit and continue the legend of the gambiered Canton gauze is really a grave issue the industry currently faces.

Another dilemma is the problem of market demand. The 1930s was the heyday of gambiered Canton gauze production. The production scale was very large, with up to ten thousand workers employed. However, with the rise of market substitutes, such as chemical fibers and textiles, the output of gambiered Canton gauze dwindled year by year. By the 1980s, many gambiered Canton gauze factories closed down or switched business. At present, there are only three professional producers of gambiered Canton gauze in the country, and they are mainly aiming at foreign markets.

Breakthrough: Improvement Propped by the Fashion Industry

Mr. Liang Zhu, who is over 80 years old, worked as an apprentice in gambiered Canton gauze factory at the age of 14, and has witnessed the ups and downs of the development of gambiered Canton gauze. In the 1990s, in order to preserve the gambiered Canton gauze craft, he bought the closed gambiered Canton gauze factory. With accumulated experience, he innovated the gauze drying process on the basis of traditional crafts, developed the process of "three washings, nine cookings, and eighteen dryings". His gambiered Canton gauze gradually won market recognition and many overseas orders.

With the forceful support from Lunjiao government agency of Shunde municipality, in 2007 the "gambiered Canton gauze drying process", that is, the "gambiered Canton gauze dyeing and finishing technique" was successfully declared a provincial intangible cultural heritage, and in 2008 it was elevated and placed on the catalog of the second batch of national intangible cultural heritage. Furthermore, there are young people willing to learn the dyeing and finishing process of gambiered Canton gauze and apprentice with Mr. Liang Zhu to pass on the mantle.

Shunde designer Qu Tingnan, one of China's top ten fashion designers, curator of the Lingnan Costume Museum and an inheritor of Guangzhou embroidery, who grew up next to a drying factory, developed a deep affection for gambiered Canton gauze. He has been committed to the development and inheritance of the gambiered Canton gauze craft. He participated in the clothing design competition organized by the United Nations with design work based on gambiered Canton gauze, and won the "World Folk Culture Award". His effort won the approval from international experts and helped the world rediscover the gambiered Canton gauze.

Further still, Qu Tingnan also asked experts to develop a variety of colors of gambiered Canton gauze, and successfully developed a red cuit silk. In 2018 in the China International Fashion Week, Qu Tingnan's Scent, Auspice, Sound and Direction theme fashion show featured high-end fashion clothes made of colorful gambiered Canton gauze fabric, which brought Chinese intangible cultural art—gambiered Canton gauze back on the world stage again.

Gambiered Canton gauze, the beautiful fabric inherited and developed by several generations or even more than a dozen generations, will serve as a time-honored business card of foreign exchanges. In the time where values such as tradition, return to nature, environment protection are cherished, it has, with its characteristics of naturalness, blazed a path of design which perfectly blends the locality of culture and the cosmopolitan nature of fashion, gained increasing popularity among Chinese and people around the world.